The Triumph of Boulton and Watt

The Pioneers of Steam

By

T. H. Marshall

British Library Cataloguing-in-Publication Data
A catalogue record for this book is available from
the British Library

The Triumph of Boulton and Watt

" Pasha. The ships of the English swarm like flies; their printed calicoes cover the whole earth, and by the side of their swords the blades of Damascus are blades of grass. All India is but an item in the Ledger-books of the Merchants, whose lumber-rooms are filled with ancient thrones! whirr! whirr! all by wheels!—whiz! whiz! all by steam! "—KINGLAKE, *Eothen*.

THERE could be no doubt left in the mind of the public by 1780 as to the immense value of the new steam-engine, at least in so far as the pumping of water was concerned. With a healthy hum of smooth-running machinery it sailed through tasks before which the old engines would have collapsed with a sob. A useful comparison can be made on the basis of the proportion between the work done and the fuel consumed. This can be measured by the number of pounds weight raised one foot by the consumption of a bushel of coal, and it is known as the " duty " of the engine. Smeaton estimated that the average duty of the atmospheric engines in the Newcastle district in 1769 was just over 5¼ millions, and the maximum then realised was just under 7¼. Watt's first engines, of the 1776 model, got up to 21¼ millions, and by 1780 he had increased this to 26⅓. Boulton was not exaggerating when he claimed that the efficiency of the steam-engine had been increased fourfold since Watt took out his patent. The reciprocating engine was now good enough to satisfy even Watt's fastidious taste. It was time to pass on to other problems. When Boulton first contemplated the idea of manufacturing engines, it was the rotary, or rotative, type that interested him most. As the owner of a factory, he could appreciate the possibilities of steam-power for the driving of factory machinery. Watt, on the contrary, having started work on the reciprocating engine for pumping water, had consistently refused to be diverted on to other lines of experiment. The engineer's love of mechanical perfection, and the Scotsman's longing for a secure, if modest, income combined to make him stick to his chosen task until it was both a technical and commercial success. When the bitter struggle was nearly over and the end seemed to be in sight, his reluctance to embark on new projects increased. He was in terror lest he might have to face the same heart-breaking anxieties all over again, and, by dividing his attention, might even lose the advantages he had already gained. Besides, he was distinctly sceptical about the prospects of finding a market for rotatives. Up till now he had been selling good engines to people who were already using bad

---

ones, and for whom the change was an urgent necessity ; with rotatives it would be a matter of persuading people to buy who had never used an engine before, and who were getting on quite well without one.

It is easy to condemn Watt for lack of enterprise and to criticise him for failing to realise the almost unlimited scope for the application of steam-power to the factory. But in Watt's days the factory was itself a rarity. Soho was unique until Wedgwood built Etruria in 1770. There had, of course, been water-mills for centuries for grinding corn, and, if you knew where to look, you could find fulling-mills, silk-mills, and even paper-mills as well. Big distilleries had been built to satisfy the abnormal passion of the day for spirits, and there were a few big breweries, sugar refineries and glass factories, some of which were suited to the application of power. But Watt did not anticipate getting anything very startling out of them ; and he was right. In the twenty-five years of the partnership they only accounted for about one-tenth of the engines sold.

Watt looked to mining and metallurgy for his principal market, with some assistance from canals and waterworks. The iron industry had been one of the earliest customers. The first engine to start working outside Soho was erected to blow the furnaces at Broseley. And it was obvious that, since the introduction of coal fuel, the iron industry was making very rapid progress. A modern observer might consider that any firm ought to be satisfied at the prospect of enjoying a monopoly in the supply of power to all the industries that work in iron and steel. But before passing judgment, he must try to imagine a world in which steel was counted among the precious metals, in which all ships were built of wood and all bridges of wood or stone ; in which the vision of an age when machines should be made of iron by machines made of iron—and so back to the first Adam of machinery—and all should be driven by power, would have been reckoned not merely among the Nightmares, but among the Revelations. When Watt came to Soho, our total output of pig-iron was only about 50,000 tons a year. A century later it was 7½ millions. The prophets of those days were expecting rapid development ; but even the most sanguine prophets, when trying to estimate human progress, hesitate to employ the 150 times table. However, the outlook here and in the mining areas was distinctly encouraging, and even though a great part of the work could be done by reciprocating engines, there would be a demand for some rotatives to drive rolling and slitting mills, polishing machines and tilt hammers, and to run the winding gear at the pits.

What was in fact to be the scene of the greatest and most immediate triumph of his engine Watt, in 1780, could not foresee. If any one had then told him that, of the 325 engines destined to be produced between 1775 and 1800, 114 would go to the textile industries, and 92 of those into cotton mills, he would have been entirely incredulous. For centuries the manufacture of woollen cloth had been carried on in the homes of the weavers and spinners. Nobody could expect any sudden change there, for habits strike deep roots in four hundred years. The excitement about cotton seemed to be merely silly, reminiscent, somewhat, of the activities of the Projectors in the seventeenth century. It was well known that for a long time cotton had been spun into thread and mixed with linen, wool, or silk, to make a variety of fabrics, and that this industry had brought prosperity to a considerable district in Lancashire. But nobody made pure cotton goods. The use of printed cottons was prohibited early in the eighteenth century, because they were all imported from India, and they injured the native woollen industry.

Then, in the very year in which Watt took out his first patent, Arkwright patented his machine for spinning with rollers. This was the first machine that could spin thread strong enough to allow the manufacture of fabrics of cotton only. The way was opened to a new industry with the prospect of fabulous profits for those first in the field. The invention was important for another reason. Machines for roller-spinning can very easily, and most advantageously, be driven by power. A horse was the particular form of power Arkwright had in mind, but he soon replaced it by a water-wheel. The water-wheel brought the spinning - mill. But Arkwright's machine alone did not transform the cotton industry. The decisive step came still later with the invention of the " Mule " and of its cousin, the " Billy," which came into use about 1780, and these also were suited to the application of power.

Up till now Arkwright's iniquitous patent for a process he had not really invented had cast rather a shadow over the industry. No one was supposed to use a spinning machine that contained rollers unless they paid him for the privilege, and rollers were used in the " Mule." But the remarkable success of Arkwright himself, and of those to whom he sold a licence to set up his machinery, proved too tempting, and in the eighties his rights were constantly being invaded. He tried to prosecute the offenders, but action at law only revealed the weakness of his case, and in 1785 his patent was finally quashed. This year marks the beginning of the real boom in the cotton industry.

The cautious mind of Watt was scornfully distrustful of this reckless behaviour. The bubble was being over-inflated and would surely burst. He had no intention of being involved in the calamity. He wrote to Boulton, who had gone to Ireland on patent business : " If you come home by way of Manchester, please not to seek for orders for cotton-mill engines, because I hear that there are so many mills erecting on powerful streams in the north of England, that the trade must soon be overdone, and consequently our labour may be lost." How could such a trade hope to have a future ? There was not in history an example of an industry of first-rate importance being established in a country which could not produce a single ounce of the necessary raw material. The thing was unthinkable. It is true that in those days the raw cotton came from our possessions in the West Indies ; but if it had continued to do so, and if we had been afraid to become dependent on the supplies grown in the United States, our manufacture of cottons would never have rivalled in importance our manufacture of woollens. Watt can be excused for looking askance at this monstrosity. He was watching the beginning of a new chapter in the economic history of the world.

Watt had but little conception of the great future that was in store for his invention. Even Boulton, who was quicker to grasp the significance of the movements of commerce, so far underestimated the coming demand for steam-engines, that he imagined that the factory at Soho would be equal to satisfying, for many years to come, the needs of the whole world. But Boulton was fully alive to the importance of getting a rotary engine put on the market as soon as possible, even if its use were to be confined to mills that were already employing water-power to drive their machinery, and he persuaded Watt to concentrate his attention on this problem.

Watt had only been deterred by his misguided ideas as to what would be profitable; all his scientific instincts urged him to explore every mechanical variation of the steam-engine. He had from the very first noted this in his mind as a problem that must some day be solved, and the moment he went seriously to work at it he became completely absorbed. The work of invention was infinitely more congenial to him than the duties of prospector and commercial traveller which he was often called on to perform.

In his original patent of 1769 Watt had included a device for obtaining a rotary motion which he generally referred to as a "steam-wheel." The wheel was hollow, and was driven round by the direct action of the steam passing within it. Nothing much had come of this idea, but both he and Boulton had played with it at intervals ever since. While he was thinking out the designs for his original steam-wheel, Watt had seen at a colliery an engine, in which the vertical motion of the rod attached to the beam was converted by a system of cogs into a rotary motion to drive a wheel. The engine was of the ordinary reciprocating type, and as it was only the upward stroke of the rod that had any driving force, the motion given to the wheel was very irregular. The machine was too clumsy to have any interest for Watt.

Some years later he saw a very similar engine, built by a certain Matthew Wasborough to drive a rolling-mill at Birmingham, but it had been greatly improved by the addition of a fly-wheel. Now Wasborough was a quite inferior engineer, and the spectacle of his apparent success wounded Watt's vanity. He was convinced that he could make a better rotary engine than Wasborough. He determined to get the circular motion by means of the common crank, and to make the motion regular by constructing an engine with two cylinders acting on two cranks attached to one axis. In this way, whenever one crank was idle, the motion was being communicated through the other. On these lines he made a model, and, as he tells us, he " employed a blackguard of the name of Cartwright (who was afterwards hanged), about this model," who went off and gave a full account of it to a large gathering in a public inn. Whereupon one of his audience hurried up to London, took out a patent for the use of the crank to obtain a rotary motion, and concluded an agreement with Wasborough for its application to his engine.

Watt was infuriated by this piece of treachery. It had never occurred to him that any one could claim to patent the crank, for, as he said to his son, "the true inventor of the crank rotative motion was the man (who unfortunately has not been deified) that first contrived the common foot-lathe." Defeat at the hands of so contemptible a rival made him bitter. "If the King should think Matt. Wasborough a better engineer than me, I should scorn to undeceive him ; I should leave that to Matthew. The conviction would be the stronger, as the evidence would be undeniable ! "

If he had challenged the patent, he could almost certainly have overthrown it, but he was afraid to create a precedent for the annihilation of patent rights for fear that he himself might be the next victim. He was forbidden to use the crank ; very well, he would do without it. He sat down and drew up plans of five alternative ways of adapting a steam-engine to drive a wheel, and sent them to Boulton. "I send you enclosed," he wrote, "three yards of the specification, and have about one yard more to send, which is the explanation of the drawings. . . . I have thought on some other methods by which rotative motions may be made, but they are inferior to those specified, and I feared the specification would have grown four yards long."

He patented the lot in 1781, but only one of them was ever used, and that only until the lapsing of his rival's patent set him free to adopt the crank. It was known as the " Sun and Planet " motion, and it has been asserted that it was originally invented by William Murdock; but Watt claimed it as an old idea of his own, " revived and executed by Mr. M." One cog-wheel is fixed to the end of the driving-rod, and works into another, attached to the axle of the wheel to be driven, in such a way that it makes two revolutions for every stroke of the engine.

Watt now got into his stride, and the flow of his ideas inundated many more yards of specification paper. It was at this point that he brought to perfection and patented, in 1782, the double-acting engine already described. It was especially suited to rotative engines, as its double stroke, upwards and downwards, solved the problem of continuous motion that had baffled Wasborough. It was a complicated machine, and therefore more liable to accidents, but workmanship was improving, and so also was the skill of the mechanics who were set to tend the engines.

The double-acting engine in its turn gave rise to a new problem. In all engines of this period, the rod of the piston was attached to the end of a beam, pivoted at its centre. The piston rod must move in a vertical straight line. If it does not, it will strain the joint where it enters the cylinder and let the steam escape. But the end of the beam moves in a curve. So long as the piston had only to pull on the beam, it could be attached by a flexible chain. But in the double-acting engine it had to push as well. There must be some rigid connection which would not wrench the piston out of the straight.

Watt's solution of this tricky little problem by means of the famous " Parallel Motion," which was patented in 1784, is, for his biographer, the most tantalising event in his life. It is the most beautiful and fascinating of his inventions, and is quite indescribable on paper, even with the help of a diagram. A parallelogram of jointed rods is fixed on the under side of the beam, and one angle is fastened to the head of the piston-rod. The whole contraption is carried through the curve described by the end of the beam, but, as it goes, its joints, obedient to the mysterious laws of geometry, perform a delicious, sinuous wriggle, and the angle fastened to the rod beats boldly up and down along a perfect vertical. Many tongues have sung the praises of this " beautiful invention." His contemporaries said that " Mr. Watt's Parallel Motion alone will immortalise his name as a mechanic." Its charm was universal, and the following account by an eminent engineer seems to have hit on the true explanation. " It is indeed impossible," he writes, " even for an eye unaccustomed to view mechanical combinations, to behold the beam of a steam-engine moving the pistons, through the instrumentality of the parallel motion, without an instinctive feeling of pleasure at the unexpected fulfilment of an end by means having so little apparent connection with it." It was, in fact, as inexplicable, as inconsequent, as spontaneous, as the works of Nature, and Watt felt a thrill of pride as he watched this creation of his genius moving in a mysterious way its wonders to perform. " Though I am not over anxious after fame," he wrote, " yet I am more proud of the parallel motion than of any other mechanical invention I have ever made."

The translation of these ideas from his brain to paper, and their embodiment in matter, involved much patient and often tiresome labour. The painful contrast between the swiftness of the conception and the slowness of the realisation brought on fresh bouts of irritable depression. " These rotatives," he said, " have taken up all my time and attention for months, so that I can scarcely say that I have done anything that can be called business. Our accounts lie miserably confused." He employed a man named Playfair to make the drawings, but they were so bad that he could not use them. " Therefore I fear I must draw the whole over myself, which, in my present state of health, and hurried as I am, is dreadful to me." He started to do it, but suffered such pain in his head and back that he nearly gave up the task. But his will to work overcame the temptation to surrender, and ten days later he wrote, " I have got one copy of the specification drawing finished in an elegant manner upon vellum, being the neatest drawing I ever made."

But while confident that he could solve the technical problems of the rotative engine, he continued to have doubts as to its commercial value. It is possible to trace his gradual conversion. " I have a very mean opinion of the rotative's profits," he wrote in January 1782, " and the trouble with each of them must be at least double that of an engine that raises water. Peace of mind, and delivery from Cornwall, is my prayer." In November of the same year we find him writing, " There is now no doubt but that fire-engines will drive mills, but I entertain some doubts whether anything is to be got by them." Two years later his tone had changed. " Our rotative engines," he writes, " are certainly very applicable to the driving of cotton mills, in every case where the conveniency of placing the mill in a town, or ready-built manufactory, will compensate for the expense of coals and of our premium." By 1786 the designs were completed, the double-acting rotary engine was a proved success, and orders were pouring in so fast that it was almost impossible to find men enough to execute them.

The capacity of this new market for engines was almost inexhaustible, and the partners knew that · henceforth they would receive as many orders for machines as they chose to undertake. And their customers were no longer, like the copper miners, a crew of beaten, broken adventurers, searching desperately for some means of checking the rot in their fortunes ; they were a company of healthy, vigorous pioneers borne on the rising tide of a new prosperity. In this prosperity Boulton and Watt could claim a share. They were now at the height of their fame. The industries of England competed for the favour of their attention.

Watt had the honour of explaining one of his engines to George III at Whitbread's brewery. " His Majesty," he says, " was much pleased with the brew-house, which is immense." Shortly afterwards he visited the King at Windsor, and was obliged to answer the intelligent questions that royalty is accustomed to ask about the activities of its subjects. In 1786 Boulton and Watt proceeded to Paris, at the invitation of the French Government, to consider the erection of a steam-engine to take the place of the famous and prodigious machine of Marly, built in 1682 to raise water to supply the town and the waterworks of Versailles. Nothing came of the proposal, but they thoroughly enjoyed their visit. The official reception was magnificent. It was the first time Watt had been treated as a "distinguished foreigner," and he was much flattered. He was "drunk from morning to night with Burgundy and undeserved praise," or so at least he says. But most gratifying of all was the welcome given him by the leading scientists of France, who treated him as an honoured colleague and flocked to hold conference with him.

Even in this time of apparent triumph Watt's letters are full of lamentation. It is not unnatural. Owing to his constitution, work of any kind was a strain, and always produced a nervous reaction ; but whereas work at his scientific experiments gave him a kind of nervous exaltation, anything of the nature of business worries or responsibilities brought on a condition of nervous exhaustion. In 1782 Boulton had handed over to him the management of the firm's accounts, and since that date Boulton had been more and more in the habit of going off on his own affairs, leaving the full responsibility for the direction of the business on Watt's shoulders. Expansion was at this time very rapid, and the burden was more than Watt could bear. He groaned under its weight, and sighed for the rest that only retirement could bring.

"I should have written to you long ago," he

writes on July 18th, 1786, " but have really been in a worse situation in some respects this spring than I have ever been in my life. The illness I was seized with in London, in the spring, greatly weakened me both in body and mind. . . . The bodily disease has in great measure subsided ; but an unusual quantity of business, which by Mr. Boulton's frequent and long absences has fallen wholly on me, and several vexations, with the consequent anxious thoughts, have hitherto prevented my mind from recovering its energy. I have been quite effete and listless, neither daring to face business, nor capable of it ; my head and memory failing me much ; my stable of hobby-horses pulled down, and the horses given to the dogs for carrion. I have had serious thoughts of throwing down the burthen I find myself unable to carry, and perhaps, if other sentiments had not been stronger, should have thought of throwing off the mortal coil ; but, if matters do not grow worse, I may perhaps stagger on. Solomon said that in the increase of knowledge there is increase of sorrow : if he had substituted *business* for *knowledge*, it would have been perfectly true."

Matters were made worse by the fact that, at this moment, Boulton was facing a financial crisis, perhaps the most serious of his life. The engine business was doing well, Watt for the first time was free from debt and had a comfortable balance at the bank. But Boulton was deeply involved in other speculations, some of them, like his investments in the copper mines, indirectly connected with engines. In 1787 trade was depressed. There had been considerable over-production in the cotton industry, and manufacturers had difficulty in disposing of their stocks. Several big London firms of merchants were involved, and in 1788 there was a crop of failures, including an old-established Manchester bank. Boulton badly needed an extension of credit, but it was extremely hard to get. He appealed to Watt for assistance. But Watt, with characteristic caution, had already safely invested his money, and the appeal was made in vain. When it is remembered that Boulton had, throughout the hard years of struggle, taken all the financial risks and worries on his own shoulders, that he had paid Watt a regular salary when the business was not making a penny, and had, out of pure generosity, allowed him half-profits, instead of the stipulated one-third, when profits began to come in, Watt's action at this crisis appears mean and ungrateful. Money matters always brought out the worst in him. His horror of the jugglings of finance, his dread of instability of income, amounted almost to a disease. To withdraw money from a safe investment and throw it into a speculative venture seemed to him not merely a pity, but a crime, a kind of child-murder. It was a crime that he could not bring himself to commit, even to help a friend.

Boulton weathered the storm, and his prosperity was never again in danger, but the strain had permanently damaged his health. Both the partners were beginning to look forward to the time when they would be able to retire from all active share in the business. Their two sons were being trained to succeed them, and by 1795 they were participating in the work of management. The partnership and the patent rights were both due to come to an end in 1800. But before this goal could be reached, there was one more battle to be fought.

There had always been trouble from pirates —men who picked up some knowledge of the principle of Watt's engines and made use of it without recognising their debt to the inventor. As a rule the machines they produced were so inefficient that it was not worth while to stop them. They were either like Hornblower's engine at Radstroke, which was "obliged to stand still every ten minutes to snore and snort," or like Evans's mill, which "was a gentlemanly mill : it would go when it had nothing to do, but refused to do any work." Occasionally excitement was provided by the bursting of a boiler, but the engines were rarely able to develop enough energy to achieve this ; for in all, as Watt quaintly expressed it, "the bodily presence was weak."

In time, however, as the machines became more familiar, and an ever-increasing number of men passed through the Soho works and went out skilled engineers, not scrupling to use their skill to defraud their late masters, piracy became a more serious matter. Firms ordered engines of Watt's design from these men, and of course paid Watt no dues on them. When the engines proved unsatisfactory, they blamed Watt, and the credit of Soho suffered. Others, who had Soho engines, refused to pay their dues, because they saw that their neighbours were using a similar machine free of charge.

In these circumstances Boulton and Watt decided to put their rights to the test of law. It was not merely the loss of revenue that disturbed them. In any case the patent had only a few years to run. Their pride was involved. If they submitted without protest, it would amount to an admission that their business was built on a fraud, that the invention was a sham, and that all the payments they had been drawing from their customers had been exacted on false pretences. The idea was intolerable. They began to prosecute the offenders. In 1793 action was taken against a man of the name of Bull, who had been employed by the firm as a stoker. The case was perfectly clear, and the jury quickly decided that the patent had been infringed; they left it, however, to be determined by a special case in the Court of Common Pleas whether the patent was in itself good and valid. This point came up for trial two years later before the Lord Chief Justice and three judges.

These learned gentlemen had little to say about steam-engines, but many profound thoughts that they were burning to deliver on the subject of the patent laws. Each in turn gave his display of rhetorical juggling, spun his argumentary hoops and jumped through them. Was the subject of the patent a process or only a principle ? And if a process, and a new process, was it based on an old principle ? Or was it again a machine, or only part of a machine ? And if a part, was it merely a new part of an old machine ? So profound was their knowledge of the law, and so complete their ignorance of the properties of steam and the history of invention, that when they came to apply their general conclusions to the particular case, the nature of the issue was a matter of pure chance. On they pounded round the circular track of their arguments, like racers in a stadium, but there was no common goal. Each one carried his own winning-post in his pocket, and erected it as soon as he began to feel tired. It was the judges, rather than the case, that were in the scales, and blind Justice secured equilibrium by putting two into each.

This divided opinion on a matter of such importance was most unsatisfactory, but it was swept away by the decisive victory in the following year in the case against Hornblower and Maberly. All Watt's old friends, with Robison at their head, rallied to his defence and routed the forces of Jabez, son of Jonathan. But that was not the end. The case was tried again on a writ of error, and it was not till 1799 that Watt could write triumphantly to Boulton, " We have WON THE CAUSE hollow. All the Judges have given their opinions very fully in our favour." Even after this the engine pirates continued their operations, but, said Watt, " having become used to them, we do not lay them so much to heart as formerly." They caused him a loss of revenue and heavy expense in legal proceedings, but the honour of the firm had been vindicated, its prestige was high, its reputation unchallenged.

When the century drew to its close, Watt was in his sixty-fourth year. A chapter in his life was ending. When, thirty-five years before, there had come to him the first inspired vision of his new steam-engine, all the labour that followed was but its necessary sequel. The invention must be perfected, manufactured, delivered to the public. Until this had been achieved the process was incomplete. So on he laboured, finding little joy in the work itself, and forcing himself to endure much that was almost intolerable, impelled still by that first desire to create which would not let him lay aside his tools until the task was done.

Now at last he had finished. His invention was as perfect as he could make it, the business was prosperous, his engines were at work in all the great industries of the country. Quietly, with no regrets but only profound satisfaction, he passed from the scene of his now completed labours, not into idleness, but to occupy himself with new thoughts and new projects as fascinating and absorbing as those of old.

Lightning Source UK Ltd.
Milton Keynes UK
UKOW04f2046170215

246460UK00001B/47/P